THE WEA...

FOG

A Crabtree Roots Book

DOUGLAS BENDER

CRABTREE
Publishing Company
www.crabtreebooks.com

School-to-Home Support for Caregivers and Teachers

This book helps children grow by letting them practice reading. Here are a few guiding questions to help the reader with building his or her comprehension skills. Possible answers appear here in red.

Before Reading:

• What do I think this book is about?
  - *I think this book is about fog.*
  - *I think this book is about where fog comes from.*

• What do I want to learn about this topic?
  - *I want to learn where fog forms.*
  - *I want to learn what fog is made of.*

During Reading:

• I wonder why…
  - *I wonder why fog forms in specific places.*
  - *I wonder why fog stays on land.*

• What have I learned so far?
  - *I have learned that fog is made of water droplets.*
  - *I have learned that fog forms around mountains.*

After Reading:

• What details did I learn about this topic?
  - *I have learned that the Sun makes fog disappear.*
  - *I have learned that fog can form in many different places.*

• Read the book again and look for the vocabulary words.
  - *I see the word **cloud** on page 4 and the word **mountains** on page 6. The other vocabulary words are found on page 14.*

This is what **fog** looks like.

Fog is a **cloud** on land.

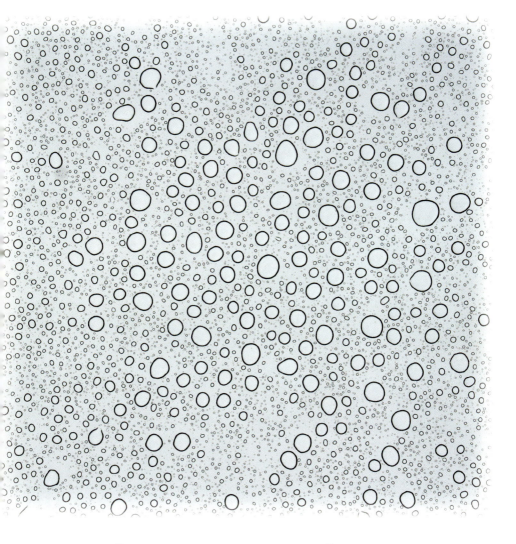

Fog is made of water **droplets** in the air.

Fog can form around **mountains**.

Fog can form over water.

Fog can form in **valleys**.

When the **Sun** comes out, fog goes away.

Word List

Sight Words

a	like	Sun
around	looks	the
away	made	this
can	of	water
goes	on	what
in	out	when
is	over	

Words to Know

cloud

droplets

fog

mountains

Sun

valley

44 Words

This is what **fog** looks like.

Fog is a **cloud** on land.

Fog is made of water **droplets** in the air.

Fog can form around **mountains**.

Fog can form over water.

Fog can form in **valleys**.

When the **Sun** comes out, fog goes away.

THE WEATHER FORECAST

FOG

Written by: Douglas Bender

Designed by: Rhea Wallace

Series Development: James Earley

Proofreader: Janine Deschenes

Educational Consultant: Marie Lemke M.Ed.

Photographs:
Shutterstock: Budimir Jevtic: cover; Mikhail Piskunov:
p. 1; Woody Alec: p. 3, 14; Alexander Steam: p. 4,
12, 14; Lillac: p. 6-7, 14; Ann Tikhonova: p. 9; Rusian
Suseynov: p. 11, 14

Library and Archives Canada Cataloguing in Publication

Title: Fog / Douglas Bender.
Names: Bender, Douglas, 1992- author.
Description: Series statement: The weather forecast |
    "A Crabtree roots book".
Identifiers: Canadiana (print) 20210181397 |
    Canadiana (ebook) 20210181419 |
    ISBN 9781427159328 (hardcover) |
    ISBN 9781427159380 (softcover) |
    ISBN 9781427133762 (HTML) |
    ISBN 9781427134363 (EPUB) |
    ISBN 9781427159564 (read-along ebook)
Subjects: LCSH: Fog—Juvenile literature.
Classification: LCC QC929.F7 B46 2022 | DDC j551.57/5—dc23

Library of Congress Cataloging-in-Publication Data

Available at the Library of Congress

Crabtree Publishing Company

www.crabtreebooks.com   1-800-387-7650

Printed in the U.S.A./062021/CG20210401

Published in the United States
Crabtree Publishing
347 Fifth Avenue, Suite 1402-145
New York, NY, 10016

Published in Canada
Crabtree Publishing
616 Welland Ave.
St. Catharines, Ontario L2M 5V6